Scholarly Reprint Series

The Scholarly Reprint Series has been
established to bring back into print
valuable titles from the University of
Toronto Press backlist for which a
small but continuing demand is known
to exist. Special techniques (including
some developed by the University of
Toronto Press Printing Department)
have made it possible to reissue these
works in uniform case bindings in runs
as short as 50 copies. The cost is not
low, but prices are far below what
would have to be charged for such
short-run reprints by normal methods.

The Scholarly Reprint Series has
proved a valuable help to scholars and
librarians, particularly those building
new collections. We invite nominations
of titles for reissue in this form, and
look forward to the day when, with this
series and other technological de-
velopments, the label 'out of print' will
virtually disappear from our backlist.

UNIVERSITY OF TORONTO
BIOLOGICAL SERIES NO. 62

Observations on the Alewife, *Pomolobus pseudoharengus* (Wilson), in fresh Water

J. J. GRAHAM

PUBLICATIONS OF THE ONTARIO FISHERIES
RESEARCH LABORATORY, NO. LXXIV

UNIVERSITY OF TORONTO PRESS : 1956

CONTENTS

ABSTRACT

THE MAJOR STUDY in this investigation was one of growth and form, carried out to evaluate differences in growth rates and body proportions between the land-locked Lake Ontario and the anadromous Atlantic alewives. Particular attention was also given to the nature of the annual mortality that is characteristic of *P. pseudoharengus* in Lake Ontario. The Atlantic alewife has a more rapid rate of growth than the landlocked form. Both groups display early rapid growth followed by a decrease in the growth rate coincident with the onset of sexual maturity. It is suggested that the freshwater environment hastens the onset of maturity with its attendant inhibition of growth. The Lake Ontario alewife matures about one year earlier than the Atlantic alewife.

Variation of form of the alewife within Lake Ontario is essentially small. Of thirteen measurements made, only three showed slight differences between samples. In contrast there was a wide difference between the form of the Lake Ontario and the Atlantic representatives. Eight of the thirteen characters compared showed significant differences in their growth partition constants. Three of the remaining five characters showed differences in the intercept of the relative growth lines.

The head length of the Lake Ontario sample was measured at three sizes representing small, medium, and large fish; the range decreased progressively with increasing body length. The slope of the line representing the relative growth of the head length in the Atlantic sample roughly paralleled the upper limit of variability as set by twice the standard deviation, but converged with the lower limit similarly set. It is concluded that the differences in form between the two groups are chiefly the result of differential selection of the freshwater alewife by its environment. Inadequacy of osmotic regulation by the Lake Ontario alewife is thought to be acting, possibly in conjunction with temperature, to remove small-headed fish from the population.

During the alewife's spring entry into shoal water in Lake Ontario it sometimes encounters gradients that have maximum surface temperatures approaching 20°C. Alewives were observed to die principally within the upper ranges of these gradients. A correlation between the gradual increase in surface temperature from spring to summer with increase in mortality was also found. Comparison of the range of lethal temperatures found in the laboratory with the lethal ranges of the temperature gradients observed in the field showed that the early incidences of mortality are related to the entrance of the alewives into warm shoal waters in spring while they are still acclimated to the low temperatures of the lake's depths. Later and more extensive mortalities are believed to be also caused by lethal temperatures although no support for this belief was obtained in the laboratory.

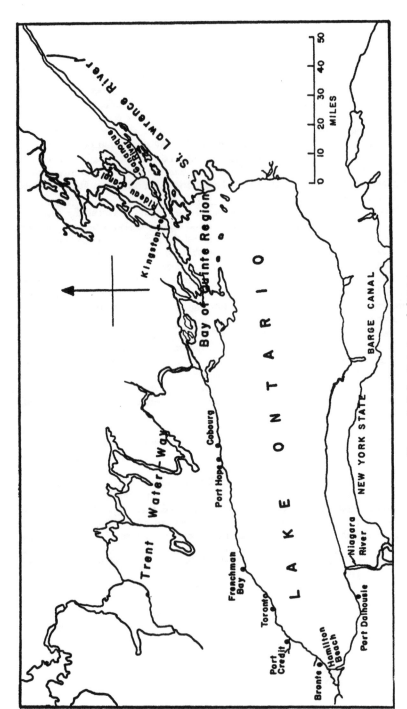

FIGURE 1. Lake Ontario and district.

Observations on the Alewife, *Pomolobus pseudoharengus* (Wilson), in Fresh Water

INTRODUCTION

THE MAJOR STUDY in the investigation was one of growth and form, carried out to evaluate differences in growth rates and body proportions between the landlocked Lake Ontario and the anadromous Atlantic alewives. Particular attention was also given to the nature of the annual mortality that is characteristic of this species in Lake Ontario.

ACKNOWLEDGMENTS

The problem was carried out under the direction of Dr. F. E. J. Fry, and I am grateful for his guidance during the investigation. I should also like to thank Dr. A. G. Huntsman of the Fisheries Research Board of Canada for his co-operation in making available both data and specimens of the Atlantic alewife. Collections of Atlantic alewives were made by Dr. V. M. Davidson and Mr. D. Alderdyce.

Field work was accomplished with the aid of commercial fishermen and resort owners, among whom I am particularly indebted to the following for services and the use of equipment—Bay of Quinte region: Mr. Donald Thurston and his son; Mr. A. Cooper; the owners of Bill and Ken's resort; the owners of Baycrest Lodge; and Mr. J. R. Tomkins of Gananoque. In the Port Credit area I am especially grateful to Mr. Edwin Joyce of Port Credit for his zealous interest and co-operation during the investigation. I am also grateful to Mr. J. N. Anderson.

The statistical procedure used in the relative growth analysis was developed under the guidance of Dr. D. B. DeLury of the Ontario Research Foundation. Dr. J. M. Speirs was very helpful in obtaining publications needed by the author and made many suggestions concerning the manuscript. I am indebted to the late Mr. Percy Ghent who X-rayed alewives used in the analysis of vertebral counts. I should also like to express my appreciation for the assistance rendered me by my colleagues of the Department of Zoology, University of Toronto.

The experimental work was carried out at the Laboratory for Ex-

[1]Taken from a thesis submitted to the University of Toronto in partial conformity with the requirements for the degree of Doctor of Philosophy. This thesis should be consulted for the original data which are tabulated therein.

perimental Limnology, Maple, Ontario, which is operated jointly by the Department of Lands and Forests and the University of Toronto. Financial support for the work was given by the University of Toronto, the National Research Council of Canada, and the Research Council of Ontario.

Materials and Methods

Growth rate

The ages of 168 specimens from the Bay of Quinte region of Lake Ontario and 141 Atlantic specimens were determined by the scale method. Most of the Bay of Quinte specimens were members of a relative growth series collected there. Thirty-four specimens of the Atlantic sample were secured from Silver Lake, New Brunswick, during the spawning run (July 14 through July 16, 1950). The remainder were captured during their downstream run in Fletcher's Creek, Shubenacadie River, Nova Scotia (young, July 26, 27 and August 2; adults, July 9, 1950).

Scales were taken from the left side in the midline above the origin of the anal fin. This area was found by Huntsman (1918) to be the first to develop scales. The scales were cleaned in water with a small brush and then observed in water under the microscope.

Relative growth

The main collections of the freshwater specimens were made in the Bay of Quinte region. But to obtain a general survey of the variability within Lake Ontario itself, samples, composed chiefly of adults, were also taken from four other areas of the lake proper: at Gananoque, Frenchman Bay, Port Credit, and Bronte. The Atlantic specimens were those referred to in the previous section.

Measurements

A steel ruler divided into half-millimetres for the first ten centimetres and into millimetres for the remaining forty centimetres, was used to measure standard lengths greater than 60 mm. Below this length and for other measurements besides standard length, needle point dividers or a caliper having a vernier reading to 0.1 mm. were used according to the size of the measurement being taken. All measurement on fish smaller than 45.0 mm. standard length were made to 0.1 mm. with the use of a binocular microscope. Measurements on fish of larger standard length were also estimated to 0.1 mm., except on specimens taken from the Gananoque River which were rounded off to the nearest 0.5 mm.

With one exception, measurements were made on fish preserved in 10 per cent formalin. Specimens from the Port Credit area were measured when fresh.

The following were the measurements made:

1. *Standard length*—Distance from the most anterior part of the head to the distal end of the hypural plate.
2. *Head length*—Distance from the anterior part of the junction of the premaxillary and maxillary bones to the most distant point of the operculum (opercular membrane excluded).
3. *Head depth*—Distance from the midline of the occiput vertically downward to the ventral contour of the head or breast.
4. *Orbital length*—Greatest distance between the orbital rims (horizontal in this case).
5. *Snout length*—The least distance between the rim of the orbit and the anterior edge of the junction of the premaxillary and maxillary.
6. *Predorsal length*—Distance from the most anterior part of the junction of the maxillary and premaxillary to the base of the first dorsal ray.
7. *Body depth*—Greatest dimension (not including structures pertaining to fins).
8. *Caudal peduncle length*—The oblique distance from the insertion of the anal base to the distal end of the hypural plate.
9. *Anal and dorsal fin bases*—Over-all length of the base.
10. *Pelvic and pectoral fin lengths*—Distance from the base of the most anterior ray to the extremity of the fin.
11. *Caudal length*—Greatest horizontal distance between the end of the hypural plate and the extremity of the pinched caudal fin.
12. *Dorsal fin length*—Distance from the origin of the fin to the extremity of the anterior lobe.

Vertebral counts were made from X-ray plates on 121 Lake Ontario alewives and 70 Atlantic alewives. A check of five specimens showed that counts upon actual dissection and from the plates were the same.

Treatment of data

The specimens measured were selected to fill logarithmic classes of standard length set up at 0.05 logarithmic intervals (DeLury, 1951). An attempt was made to fill each class with measurements from ten specimens. A 5:5 sex ratio was striven for with classes of mature fish.

The division into logarithmic classes simplified treatment of data. Means of the logarithms were obtained from the original measurements for each class. These were then plotted as a regression of the logarithm of a given body part on the logarithm of the standard length (Huxley, 1932) and the line of best fit was calculated.

Tests for statistical differences between the slopes of the characters of the two groups were made by the method of covariance. Intercepts were tested when difference between slopes was not statistically significant. Only those points representing ten specimens were used in statistical analysis.

LABORATORY INVESTIGATION

Transport of alewives

In any operation of transporting or collecting, the alewives were never removed from the water. When collecting a large enclosure was formed with the net and the alewives were then bailed from it. A splashless carrying tank (Fry, 1951) was used in carrying alewives from the sampling area to the laboratory. When alewives were carried short distances about the laboratory, beakers or glass battery jars were used. Fish undergoing acclimation were never forced into a container but allowed to swim into it undisturbed. They were then transferred into a large bucket.

In spite of extreme care in handling, carrying mortalities were considerable. Of 576 adult alewives transported from the Port Credit area between June 16 and July 29, 1950, 51 per cent died in transit. Young of the year were transported between September 26 and October 6, 1950, also from the Port Credit area. At this time, 1,560 young of the year were obtained of which 45 per cent died in transit. It is thought that many of the carrying mortalities were due to over-exertion resulting from fright (Huntsman, 1938).

Maintenance of samples

A wide variety of foods were used, but *Daphnia* and suspensions of a mixture of strained liver, entrails, and solidified blood were staples. These two foods were alternated in daily feedings. When samples of fish were small, feedings consisted entirely of live food, generally *Daphnia*.

Alewives were fed six hours before lethal temperature experiments were begun. If the experiments lasted beyond one day, the specimens were fed *Daphnia* every 24 hours.

The mortalities recorded from the holding tanks followed a definite pattern. The transfer of the sample from the lake to the holding tanks was followed by a very high initial mortality lasting a few days. This high mortality is probably the result of handling and the conditions of transport. A reduction in mortality followed, which in one case approached the low average value of 1 per cent per day. Towards the end of this period of apparent equilibrium, a noticeable increase in mortality generally occurred. Efforts were made to use a sample before this increase began. Thirty-two days was the longest period that a sample was maintained, although individual fish sometimes lived for about two months.

Disease was infrequent in the acclimating fish with the exception of *Saprolegnia*. Treatment with Roccal caused a noticeable mortality even though dilutions of 1:140,000 with exposures of approximately 30 minutes were used. Any fish showing fungus was removed from the sample and in one case an entire sample was discarded. The growth of the fungus appeared to be correlated with the amount of waste food present and with low temperatures.

Thermal acclimation

Efforts were made to acclimate alewives long enough to remove possible influences of their past thermal history. If the number in a sample decreased to as low as 20, the sample was used regardless of its period of acclimation in the laboratory. In such emergencies due consideration was given to the probable thermal history of the fish just prior to capture.

Adults and under-yearlings were acclimated in large and medium holding tanks having capacities of 75 and 48 gallons respectively. These were supplied with running well-water appropriately heated or cooled. The temperatures in these baths were maintained constant, within 0.5°C.

Acclimation of adult alewives to temperatures of 10°C, 15°C, and 20°C was attempted. Probably the so-called 10°C acclimated fish were actually adjusted to temperatures between 10°C and 15°C because of the limited time available to acclimate them. Acclimation to 15°C and 20°C is thought to be accurate because these temperatures were closer to those experienced by the samples in their lake environment before capture. Under-yearlings were acclimated to temperatures of 5°C and 9°C. A shorter range of acclimation temperatures (4 degrees), a close correlation of these temperatures to those of the lake environment, and a longer period of acclimation have made greater accuracy possible in lethal experiments involving under-yearlings.

Acclimation to low temperatures (5°C) was carried out by progressive lowering of temperature 0.5°C to 1°C every four days. Acclimation to higher temperatures (15°C and 20°C) was approached at approximately 1°C per day.

Lethal temperature experiments

The baths used in lethal temperature experiments were those designed by Brett (1952) and measured 22 inches square and 11 inches deep. Their inner galvanized surfaces were painted with a plastic paint (Tygon). The maintenance of the desired lethal temperature was achieved in each bath by 120 watt heaters, thermostatically controlled, which balanced a slow flow of water (9°C to 10°C). A three-foot coil of vinylite tubing of ⅜-inch diameter with an adjustable flow of cold water was placed in the bath when lethal temperatures were below room temperature. Generally, variation of temperature within the baths was of the order of ±0.1°C. If this variation was exceeded, the experiment was either abandoned or continued with reservations about the following results. Air saturation within the baths was found to be 70 per cent at temperatures above 26.0°C and was presumably greater at lower temperatures.

Five fish were employed in each test. Cold water introduced into the lethal baths with the alewives was immediately counteracted by the addition of warm and the desired lethal temperature was regained within 1¼ minutes. The time of death was recorded for each fish. Experiments were examined every hour during the first 12 hours and every four hours during the next 24. Half the interval of time from the last observation was recorded when death occurred during periods when no observations were made, for example, overnight.

The median resistance time of each experiment was obtained by plotting the order of death along a vertical axis of probability units and the time to death expressed in minutes along a horizontal logarithmic axis. The time at which 50 per cent of the sample had succumbed to the lethal temperature was read directly from the resulting graph. This method of determination has been shown to be satisfactory in previous lethal temperature experiments with fish (Brett, 1952; Fry, et al., 1946). The median resistance times were then plotted along a vertical axis denoting the temperature at which the experiments were carried out and along a horizontal logarithmic axis of time.

RESULTS

Summary of migrations and movements

Observations and the collection of data on the migrations and movements of the alewife while in the field were subordinated to the collection of alewives for relative growth analysis and the investigation of mortality, which form the major portion of this thesis. However, data obtained through personal observation and interviews with commercial fishermen are sufficient to present a general summary of the alewife's migrations and movements in Lake Ontario.

An inshore movement of adult alewives from deep water begins some time in April, but the precise time differs in various areas. The greatest number arrive inshore about mid-June in the Port Credit area and about late June in the Bay of Quinte region. The migration begun in April may last into mid-July in the Port Credit area and late July in the Bay of Quinte region. Once inshore, the alewives first appear in shallow water during the daylight hours. They are then found in shallow water at dusk and at dark and are observed in the offshore surface water during the day. During the inshore migration, adult alewives ascend rivers but the ascent is limited by dams.

Samples of early migrants are predominantly females. Samples taken from the bulk of the population, which arrives later, are predominantly males. Spawning begins shortly after the arrival of the first schools and reaches its height about mid-June in the Port Credit area and late June or early July in the Bay of Quinte region. Spawning takes place at night and is carried out in groups of three or in pairs. It is probable that the adult alewives reach the stage of ripeness abruptly, perhaps during a day.

The adults leave the inshore waters immediately after spawning. The majority of them migrate to the deep water of Lake Ontario proper some time in late August. They may leave abruptly as in the Port Credit area or gradually as in the protected Bay of Quinte and its associated reaches. They begin to appear in numbers in the deep water (150–300 feet) about mid-September, reaching abundance there by December and remaining until March.

Young of the year remain on the spawning grounds until at least the late larval stage is reached. Those in the lake proper probably collect later in protected areas, such as the mouth of the Credit River, some time in early September. They then migrate to the shallow water along the exposed shore of the lake about late October. The young of the year from the Bay of Quinte do not appear off the exposed shores until some time after December. Commercial fishermen have suggested January and February. Their migration probably ends in the shallow water of the exposed shores of the Bay of Quinte region, for example, Prince Edward Bay. Some young of the year may not migrate but remain in the Bay of Quinte proper.

The spring inshore migration of juvenile alewives is similar to that of the adults. Once inshore, they engage in a diurnal movement. They are found in shallow water at dusk (8:00 P.M.), and at dark. During the day the juveniles are located on bottom in 6 to 10 feet of water. Their distribution in the Bay of Quinte region suggests that they prefer exposed shorelines. It is likely that the time of migration of the juvenile alewives to deep water and the time spent there is similar to times of the adult alewives.

Growth rate

The mean age—length relations of the samples examined from Lake Ontario and the Atlantic coast are given in Table I. The Lake Ontario alewife displays an early rapid growth which decreases with the onset of maturity at ages II (males) or III (females). The female attains a greater size than the male and possibly has a longer life span. Female alewives have also been found to exhibit a faster growth rate than the males by the following authors: Pritchard, 1929 (Lake Ontario); Odell and Eaton, 1940 (Lamoka Lake and Kensington Reservoir). The author does not have sufficient data to compare the growth rate of the sexes of the Atlantic alewife but Rounsefell and Stringer (1943) re-

TABLE I

MEAN STANDARD LENGTHS IN RELATION TO AGE AND SEX FOR LAKE ONTARIO AND ATLANTIC ALEWIVES

(The age designation indicates the number of completed annuli, the number in brackets is the number of specimens examined. The sexes of immature and spent individuals were not determined)

Age group	Lake Ontario			Atlantic		
	Immature	Male	Female	Immature	Male	Female
0	32.1(43)	—	—	45.4(84)	—	—
I	62.0(36)	95.5 (1)	—	—	—	—
II	95.1 (4)	108.6(18)	110.0 (5)	—	—	—
III	—	128.1(11)	125.4(17)	(Spent) 147.4 (6)	—	
IV	—	139.0 (2)	141.2 (4)	170.5 (7)	175.1(11)	167.3 (9)
V	—	128.5 (2)	153.4(10)	220.5 (5)	188.6 (2)	202.5 (2)
VI	—	—	157.0 (5)	214.5 (2)	—	248.0 (2)
VII	—	—	164 (1)	—	—	—
VIII	—	—	172 (1)	—	—	—
IX	—	—	—	—	—	270 (1)

ported that the Atlantic female alewives averaged 6 to 9 mm. longer than the males in the Newmasket, Demariscotta, and Orland Rivers of Maine. As in the Lake Ontario alewife, the male appears to become sexually mature one year earlier than the female. Odell and Eaton

(1940), similarly, reported that the female alewife had a faster growth rate than the male Atlantic alewives from the Hudson River, New York. Atlantic alewives of both sexes mature one year later than Lake Ontario alewives.

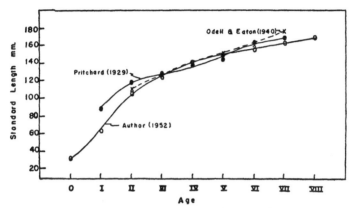

FIGURE 2.—Comparison of the growth rate of the Lake Ontario alewife as determined by Pritchard (1929), Odell and Eaton (1940), and the author. Data combined for both sexes and where necessary converted from fork or total length to standard length.

A comparison of the Lake Ontario age and growth studies by Pritchard (1929), Odell and Eaton (1940), and the author is presented in figure 2. The data of the three studies agree closely excepting for ages I and II. Possibly the disagreement for ages I and II of Pritchard and of the author is because of net selection. The divergence between the data of Odell and Eaton (1940) and those of the author for year VII cannot be considered important because of the small number of specimens.

Further comparisons of the growth of the Lake Ontario alewife with growth in other lakes and in the ocean are made in figure 3. There it will be seen that the Kensico Reservoir alewives (Breder and Negrelli, 1936; Odell and Eaton, 1940) grow more slowly than the Lake Ontario and Lake Seneca alewives. The difference between the data for Kensico Reservoir of Breder and Negrelli (1936) and Odell and Eaton (1940) is probably due to selection in the material of the former authors. Breder and Negrelli obtained their data from specimens taken from water intake screens of the reservoir during a mortality of alewives. The Lamoka Lake data are of little significance because of the small number (21) of specimens measured. The Atlantic alewife has a more rapid rate of growth throughout most of its life than the Lake Ontario alewife. The mean standard lengths of Hudson River (Atlantic) alewives are also included (Odell and Eaton, 1940). This sample shows a still higher growth rate than do the Nova Scotia specimens.

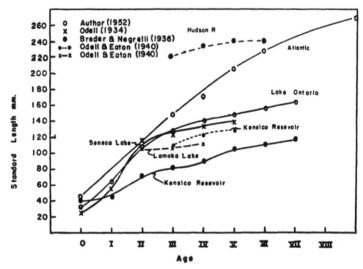

FIGURE 3.—Comparison of the growth of anadromous and landlocked alewives, sexes combined. Anadromous groups are the Hudson River (Odell and Eaton, 1940) and the "Atlantic" data of the author. The remaining curves are for land-locked populations. Data converted to standard lengths where necessary.

Relative growth

The data gathered to describe the relative growth of the alewife are presented on double logarithmic grids after the method of Huxley (1932). The slope of the line (k) represents the growth partition constant in Huxley's formula,

$$y = bx^k$$

where x is the body size, b is the intercept of the line and y is the body part measured. The resulting relative growth lines for each character are described in the terminology introduced by Huxley, Needham, and Lerner (1941) and applied to fish by Martin (1949). Essentially, the term "auxesis" is used to distinguish ontogenetic from phylogenetic relative growth series. "Bradyauxesis" is used when a body part grows relatively more slowly ($k<1$) than the body. "Tachyauxesis" is the term used when the body part grows relatively faster than the body ($k>1$) and "isauxesis" when the growth rates of the body parts and body are equal ($k=1$).

Variation within Lake Ontario

The major series of specimens used for the relative growth analysis was taken from samples collected in the Bay of Quinte region, but minor samples were also collected from Gananoque, Frenchman Bay, Port Credit, and Bronte areas (see fig. 1). The minor series were compared with the Bay of Quinte series to assess variability within Lake Ontario. The most complete comparison made was for head length; this is presented in figure 4.

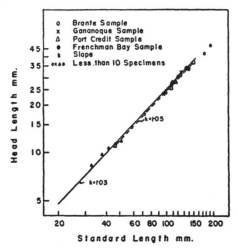

FIGURE 4.—Comparison of head lengths between the Bay of Quinte sample and minor samples taken elsewhere in Lake Ontario. The Bay of Quinte line (solid) was obtained from figure 6.

Figure 4 shows the log head length–log standard length relation for the minor series in comparison with the calculated line for the Bay of Quinte series. The Bay of Quinte line is the continuous line with a slope of $k=1.03$; the points on which this line is based are shown in figure 6. It can be seen from figure 4 that all of the points for the various series fall closely about the Bay of Quinte line. The slope ($k=1.05$) and the intercept of the line for the points outside the Bay of Quinte do not differ significantly from those of the Bay of Quinte sample.

Comparisons of the remaining characters investigated are presented in figure 5. The slopes of the lines drawn through the points are those determined statistically for the Bay of Quinte series shown in figures 6 to 11. The samples used in these comparisons are not as extensive as those used for head length, but they are made up of adult fish and conform to a similar pattern. The points for most of the characters shown in the lower half of figure 5 fall closely along the calculated lines of the Bay of Quinte sample with the exception of single points of the Bronte sample representing body depth and pelvic length. But those shown in the upper right of the figure do not agree fully with the calculated Bay of Quinte lines. The points showing the caudal peduncle length–body length relation of the Gananoque sample are more adequately described by a line (dotted) drawn parallel to, but

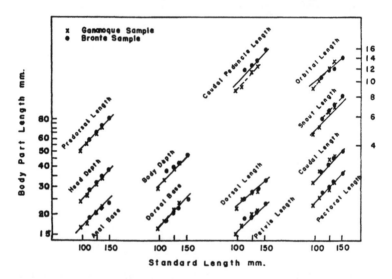

FIGURE 5.—Comparison of the variation in relative growth of various body characters of the adult alewife in Lake Ontario. The solid lines drawn through the points are those determined statistically for the Bay of Quinte series (see figures 6 to 11). The broken lines were drawn by eye.

lying a short distance from, the Bay of Quinte line. Thus, at any given standard length applicable to the above plot the Bay of Quinte sample has longer caudal peduncles than the Gananoque sample. Snout length exhibits a similar phenomenon, but here the Bronte rather than the Gananoque sample is more adequately described by a line (dotted) drawn parallel to, but lying a short distance from, that of the Bay of Quinte sample. This means that fish from the Bronte area have longer snouts than those from the Bay of Quinte. In contrast, the orbital length–standard length relationship of the Gananoque sample is described by a line (dotted) crossing over that of the Bay of Quinte sample and exhibiting a different slope. Thus the smaller Gananoque fish, described in the above plot, have a smaller orbit length, the medium-sized a similar orbit length, and the larger a larger orbit length when compared to the Bay of Quinte fish.

The parallelism found in caudal peduncle length and snout length has been shown to be the usual condition encountered in connection with variability in form between local populations of the same species (Martin, 1949; Svårdson, 1950; Hart, 1951). The results of these three authors agree in showing that such intraspecific variability is displayed by the characters following relative growth lines which are parallel to each other, but which may be some distance apart. Differences between slopes, on the other hand, are commonly considered to represent interspecific variability in form (Martin, 1949) and were found to be the main source of differences between the relative growth of the Bay of Quinte and Atlantic samples.

Comparison of Bay of Quinte and Atlantic samples

Relative growth lines for the Bay of Quinte and the Atlantic samples are combined in pairs in figures 6 to 11. The Bay of Quinte series includes fish from 17.1 mm. to 173 mm. while the Atlantic fish range from 26 mm. to 270 mm. Thus the two overlap only from 26 mm. to 173 mm. and it is within this range that they may be compared.

The presence of a "growth stanza" in certain characters of the Bay of Quinte sample, which is not represented in the Atlantic sample, also complicates the comparison. The relative growth curve for a given character of fish from hatching to senescence is rarely a single straight line on a double logarithmic grid. The pattern usually shown is a series of one or more straight lines of different slope which change sharply at certain sizes in a given population (Martin, 1949). Each portion of the curve over which a given growth partition constant holds is termed a "growth stanza". Occasionally (Thompson, 1934) the change from one stanza to another is a gradual change of slope rather than an abrupt one, but such cases have been rare to date and none has been found in the alewife in the present investigation.

Head region. Relative growth lines for the head length in the freshwater and Atlantic samples are presented in figure 6. The head length–body length relation of the alewife displays a single stanza over the size range studied. The line for the Bay of Quinte sample shows slight tachyauxesis ($k=1.03$) while that for the Atlantic sample is slightly bradyauxetic ($k=0.93$). There is a significant difference between these slopes ($P=.01$).

The comparisons of other measurements taken in the head region to body length are shown in figure 7. Two of these characters show more than one growth stanza. These are head depth and orbital length. Both initially show strong tachyauxesis ($k=1.58$ and $k=1.56$) in the early growth of the Bay of Quinte series. This is followed by an inflection at 31.7 mm. standard length in both characters. The slopes

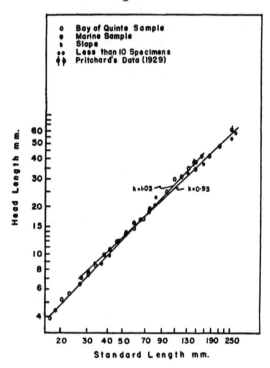

FIGURE 6.—Comparison of the head length in relation to body length between the Bay of Quinte and Atlantic samples. Double logarithmic grid. Pritchard's (1929) data from Port Credit (Lake Ontario) and Halifax Harbour (Atlantic) are also shown. In this and following figures the sexes are combined since no significant difference was found between the relative growth of the head in males and females. The data separated as to sex are on file at the University of Toronto.

following these inflections approach isauxesis ($k=1.06$ and $k=0.98$). These inflections are not present in the Atlantic sample since the smallest fish measured was 26 mm. standard length and there was little opportunity for the first stanza to appear. Presumably, this inflection occurs at a smaller standard length in the Atlantic group than in the Bay of Quinte alewife. In orbital length, there is a second inflection at about 115 mm. in the Bay of Quinte sample and at 142 mm. in the Atlantic sample.

Within the main relative growth stanza, head depth approaches isauxesis. There is no significant difference between either the slopes of the freshwater ($k=1.06$) and Atlantic ($k=1.02$) samples or their intercepts. But, there is a significant difference between slopes in the

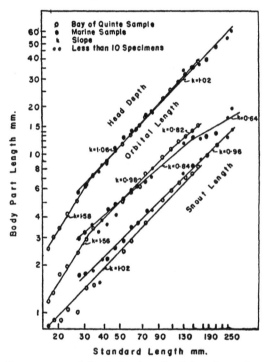

FIGURE 7.—Comparisons of head depth, orbital length, and snout length to body length for the Bay of Quinte and Atlantic alewives. Double logarithmic grid.

case of orbital length ($P=.05$) with the slope for the freshwater sample ($k=0.98$) being steeper than that for the Atlantic sample ($k=0.84$).

The snout length measurement does not exhibit any inflections, but shows only a single straight line in both the Bay of Quinte and the Atlantic samples. Two points fall well off the lower end of the Bay of Quinte line. However, no measurements made on fish smaller than 28.0 mm. standard length were used in statistical analysis. The Bay of Quinte line has a slope of 1.02 and the Atlantic line a slope of 0.96. The difference between these slopes is not significant, but that between their intercepts is significant ($P=.01$).

Trunk region. The relation of measurements taken in the trunk region to body length is presented in figure 8. The measurements made were of predorsal length, body depth, and caudal peduncle length.

All three characters show inflections of the relative growth lines at one point or another. In predorsal length, there is a stanza of extreme bradyauxesis ($k=0.40$) below 26 mm. body length in the Bay of

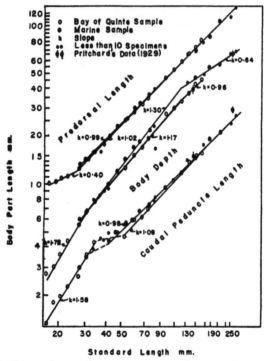

FIGURE 8.—Comparisons of measurements taken in the trunk region to body length for the Bay of Quinte and Atlantic alewives. Double logarithmic grid.

Quinte sample. Two inflections are seen in the body depth lines. The first one occurs at 30 mm. in the freshwater sample. The size range of the Atlantic sample is not sufficient to determine whether a similar early inflection occurs in this group also although the presumption is that it does. The second inflection is best displayed in the Atlantic sample where it occurs at 118 mm.; it is so weakly expressed in the freshwater group that it is impossible to fix the point of inflection with any accuracy. The slope ($k=0.96$) was obtained for the last six points which include only adult fish. The curves for caudal peduncle length were the most complex found in the investigation. There is an initial period of tachyauxesis, a pause in the growth of this region of the body and then a phase of isauxetic growth.

With respect to predorsal length there is a statistically significant difference of slope between the freshwater ($k=1.02$) and Atlantic ($k=0.99$) samples over the ranges where they can be compared

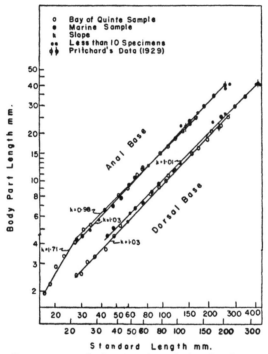

FIGURE 9.—Comparisons of the lengths of the fin bases in relation to body length for the Bay of Quinte and Atlantic alewives. Double logarithmic grid.

($P=.05$). In the case of body depth the difference in the slopes displayed in the middle growth stanza (Atlantic $k=1.30$, freshwater $k=1.17$) is highly significant ($P=.01$). It is interesting to note that the difference in body proportions brought about by this difference in slope is not maintained throughout all of the subsequent size ranges: there is convergence after the next inflection so that at 210 mm. standard length the body depth would again be the same in the two populations.

There is also a significant difference between the slopes of the two samples in the main growth stanza for the caudal peduncle length (freshwater $k=1.08$, Atlantic $k=0.98$; $P=.05$).

Fin bases. Figure 9 shows that the anal base of the young specimens of the Bay of Quinte sample grows at a much faster rate than body length, thus exhibiting strong tachyauxesis (k=1.71). Subsequent to this early tachyauxesis, an inflection occurs at about 27 mm. standard length and a shift to slight bradyauxesis takes place. A comparison of the slopes of the Atlantic line, which exhibits tachyauxesis, with that of the Bay of Quinte line above the inflection show them to be significantly different (P=.05).

The relative growth lines for the dorsal base of both the Bay of Quinte and Atlantic samples approach isauxesis (k=1.03 and k=1.01) and no inflections are present. The lines are parallel, but the difference between their intercepts was highly significant (P=.01).

Paired fins. The relative growth of the pectoral fins (length) of the Bay of Quinte and Atlantic samples is shown in figure 10. Again the Bay of Quinte sample exhibits an early fast growth of the body part concerned followed by an inflection at approximately 30 mm. standard

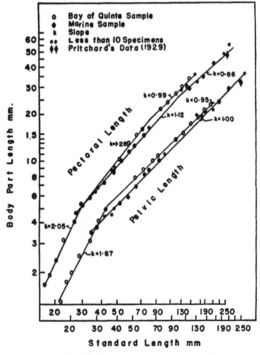

FIGURE 10.—Comparisons of the lengths of the paired fins in relation to body length for the Bay of Quinte and Atlantic alewives. Double logarithmic grid.

length. The Atlantic sample might also experience an inflection, but it would occur at a smaller standard length than is represented in the fish measured. Subsequent to the early inflection, the Bay of Quinte sample shows a decrease in the rate of growth of the pectoral fin in relation to body growth although the relation still remains tachyauxetic ($k=1.28$). The Atlantic sample shows essentially the same relative growth rate of the pectoral fin at this position in relation to body length as does the Bay of Quinte sample, since comparison of the slopes ($k=1.28$ and $k=1.12$) showed that they were not significantly different. However, a comparison of intercepts showed a highly significant difference ($P=.01$). A subsequent inflection towards bradyauxesis of the relative growth lines of both the Bay of Quinte and the Atlantic samples occurs. The Bay of Quinte line inflects at a smaller standard length (80 mm.) than the Atlantic line (126 mm.).

Figure 10 shows that the pelvic fin grows in length at a much faster rate than the body during the early growth of the Bay of

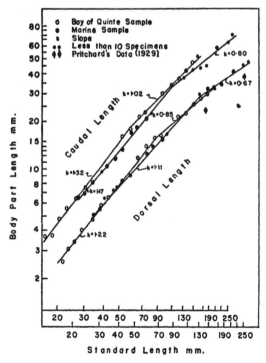

FIGURE 11.—Comparisons of the lengths of the median fins in relation to body length for the Bay of Quinte and Atlantic alewives. Double logarithmic grid.

Quinte sample. This is followed by an inflection at about 34 mm. standard length. Possibly, the Atlantic sample may experience a similar inflection but it would occur at a standard length of 27 mm. or less. Following the inflection at 34 mm. standard length, the pelvic fin of the Bay of Quinte sample grows at a slightly slower rate than the body ($k=0.95$). The pelvic fin of the Atlantic sample grows at the same rate as the body ($k=1.00$). No significant differences were found between these slopes or between the intercepts.

Median fins. No early inflection of the caudal length relative growth lines in either the Bay of Quinte or the Atlantic samples is shown in figure 11. The difference between the slope ($k=1.32$) of the Bay of Quinte relative growth line and that ($k=1.17$) of the Atlantic line was significant ($P=.05$). A subsequent inflection in the Atlantic line caused a shift from tachyauxesis to bradyauxesis ($k=0.80$) while in the case of the Bay of Quinte line the shift was from tachyauxesis to near isauxesis. The inflection occurred at a smaller standard length in the Bay of Quinte sample (60 mm.) than in the Atlantic sample (110 mm.).

Similarly, the relative growth lines for dorsal length for the Bay of Quinte and Atlantic samples show no early inflection. The difference between slopes of the Bay of Quinte sample ($k=1.22$) and of the Atlantic sample ($k=1.11$) was significant ($P=.05$). Both lines exhibit tachyauxesis, but their slopes changed to bradyauxesis following an inflection at about 68 mm. and 132 mm. standard lengths of the Bay of Quinte and Atlantic samples respectively.

TABLE II

STATISTICAL ANALYSIS OF VERTEBRAL COUNTS OF LAKE ONTARIO
AND ATLANTIC ALEWIVES

Source	Sample size	Mean and standard error
Bay of Quinte	121	47.88 ± 0.109
Atlantic (Graham)	70	48.68 ± 0.190
Atlantic (Leim)	99	48.65 ± 0.098

Vertebral counts. Vertebral counts made in connection with the above relative growth analyses were treated statistically and summarized in Table II. A comparison of the author's vertebral counts with those of Leim (1924) for 99 Atlantic alewives taken from Scotsman Bay, Nova Scotia, is included. The mean difference between the Lake Ontario alewives and both Atlantic samples (the author's and

Leim's) was statistically significant. The means of the two Atlantic samples were within the standards accepted for random variation.

The Atlantic sample had a larger number of vertebrae than the Lake Ontario sample and this agrees with comparisons of other meristic parts made by Pritchard (1929). Pritchard did meristic counts on ten specimens from the Port Credit area and ten from Halifax Harbour, Nova Scotia. He found that the Atlantic specimens had more scales, gillrakers, scutes and fin rays than specimens from the Port Credit area.

MORTALITIES

Mortalities of the freshwater alewife are common and occur at different times of the year and in various situations. Records of mortalities extend as far back as 1890 (Smith, 1892). These records, with few exceptions (Miller, 1930), are not accompanied by data regarding the circumstances of the mortalities or possible cause of death.

Annual mortality

In Lake Ontario the most spectacular of the mortalities among alewives is the "annual mortality," so called because of its almost regular recurrence. The extent of these mortalities is indicated by an estimate made July 3 and 4, 1928, of a little over a quarter of a million dead alewives on the beaches between Kew Beach, Toronto, and one mile east of Dalhousie Beach, a distance of approximately fifty miles (Miller, 1930).

The annual mortality usually occurs some time during the months of May, June, and the first half of July. The pattern of mortality begins with small incidences of death (about one thousand fish). These are followed by larger mortalities which reach their maximum in late June or early July. This pattern remains constant although the period of mortality may shift from year to year within the months of April through August (see Rathbun, 1895). Miller (1930) recorded the beginning of epidemic conditions in 1928 as June 13 in Toronto Bay and June 1 in the Bay of Quinte area. The termination of the annual mortality of 1928 was about July 12. During 1950, the first incidence of the annual mortality in the Bay of Quinte area was noted on May 26 and the last between July 7 and 12. During 1951, the first mortality occurred May 17 and the last on May 22 in the Port Credit area.

No satisfactory explanation of the mortality has been given. Miller (1930) investigated the annual mortality of 1928 and concluded that neither infections nor parasites were the cause of death. The following observations and data obtained in the present study suggest that the mortality is related to the alewife's inability to adjust to abrupt changes of temperature.

Bay of Quinte region (1950). On May 26 many dead alewives were observed on the south shore of Point Pleasant along the stretch of shoreline indicated as the mortality region in figure 12. This mortality was considered an incident of the annual mortality because of the time of its occurrence, the large number of fish involved (about 1,500 fish within a mile and three-quarters of shoreline), and the lack of apparent cause. Forty-one dead alewives were counted along 20 yards of beach at 9:30 A.M. By 1:00 P.M. the number of dead alewives had doubled in some places and in the vicinity indicated by a cross on figure 12, had formed a windrow about 50 yards long.

About 1:30 P.M. observations were begun upon a large school of alewives (at least 500 fish) which frequented the above vicinity. The school had three other species in association with it: perch, *Perca flavescens*; another species which might have been the lake emerald shiner, *Notropis atherinoides*; and a third believed to be the spottail shiner, *N. hudsonius*. These species were not affected by the mortality. The perch were preying upon the dying alewives. The movement of the school was languid and wandering. Two of the alewives had growths of *Saprolegnia* on them. All sizes (about 80 mm. to 145 mm. standard length) of alewives in the school were affected by the mortality. The stricken members of the school exhibited the following symptoms from beginning to the end of their death struggle:

1. Swimming quickly along the bottom often on their sides.
2. Increase in the rate of swimming.
3. Loss of balance accompanied by sudden darts of speed.
4. Approaching the surface, swimming in a corkscrew fashion.
5. Attempts to reach the bottom and even to burrow beneath stones
6. Swimming in circles at the surface.
7. Fluttering at the surface just before death.

This description is similar to that given by Miller (1930) for fish which died during the annual mortality of 1928. The surface temperature of the water was 19.5°C in the vicinity of point X, the wind was very slight, and the water calm. The school frequented the above position until 4:30 P.M. The surface temperature had risen to 21°C by 7:00 P.M.

A survey was carried out May 27, 10:30 A.M. to 3:55 P.M., to determine limits of the mortality along the shoreline. The gradient of temperature, shown in figure 12, was found; the higher values were essentially correlated with the areas of mortality as determined by beach counts. The mortality was greatest at the readings of 19°C and the second consecutive reading of 18.5°C. Observation in the water was hindered by a cloudy sky and waters roughened by a southwest wind and thus it was difficult to observe any dying fish. Two hours of seining with a fifteen-foot common sense minnow seine produced only five juvenile alewives. The scarcity of alewives in the area is probably related to their avoidance of rough water. Many of the dead alewives along the beach were fresh and a few were floating in from a short distance offshore, but no evidence of a recurrence of the mortality of May 26 was found. The mortality had probably begun the morning of May 26 and ended no later than noon of May 27.

The subsequent breakdown of the gradient was shown by three temperatures which were taken within the vicinity of the mortality area on May 28 and were as follows: 11°C at the position marked by a cross in figure 12, 14°C at the position registering 19°C, May 27, and 15°C at the other end of the gradient observed the previous day. The wind was south and of sufficient strength to form white-caps upon the waves.

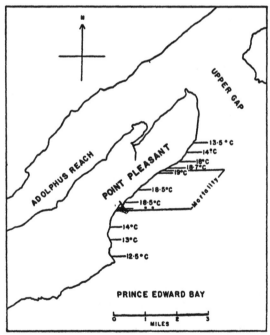

FIGURE 12.—Part of the Bay of Quinte region showing the location of a mortality of alewives on May 26 and 27, 1950, and its association with a temperature gradient.

The potentialities for the formation of the temperature gradient observed May 27 lie in the existence of shallow gravel bars and sandy shoal areas off the south shore of Point Pleasant. One of the shallowest bars is indicated on figure 12 as a stippled area. Surface water gently driven by a wind (south or southeast) apparently moved across the bar where warming of the water by the sun took place. This bar was less than 6 feet below the surface 50 yards from shore. The solar radiation necessary for providing the heat needed to warm the surface water was present during the days concerned (see figure 13). The temperature along the Adolphus Reach shoreline where such conditions are not present was 15.7°C on May 27 as opposed to the high temperatures found on the south shore where the bars are present.

Another mortality occurred in the Bay of Quinte region at a strip of beach bordering both Lake Ontario and Spencer (East) Lake in Prince Edward County. The incident took place on the Lake Ontario side

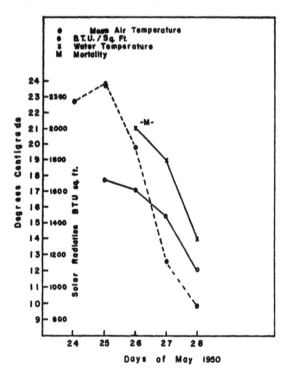

FIGURE 13.—Comparison of solar radiation and maximum air temperature recorded at Toronto with surface water temperature associated with the mortality mapped in figure 12.

and apparently resulted in the death of all the alewives in the area. The mortality was reported to have occurred some time between July 7 and 12 of 1950. On July 12 the entire beach was covered by dead young and adult alewives. No alewives were captured by seining in the mortality area and their noisy performances accompanying feeding and spawning at night had ceased.

No water temperatures were available during this incident, but solar radiation shows a pattern similar to that at the time of the incident at Point Pleasant (see figure 14).

Port Credit area (1951). Further mortalities indicating a correlation of death with temperature gradients were observed in the Port Credit area in May 1951. A temperature gradient was found in the surface water extending along 2,000 feet of the Credit River mouth and the lake shore (see figure 15). The first mortality occurred on May 17 when the surface temperature gradient extended from 9°C at the lake shore to 19.5°C within the river mouth. The high surface temperature of the river mouth was preceded by four days of increased solar radiation. The mortality subsided by May 18 when only one fish was observed to be in difficulty. Dying fish were observed again on May 22 when the surface water temperature of the river mouth, after three days of increased solar radiation, rose to 19.8°C. In both instances of mortality,

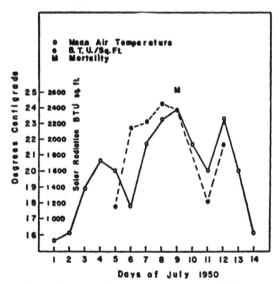

FIGURE 14.—Solar radiation and air temperature recorded at Toronto over the period at which a mortality occurred in 1950 in the vicinity of Spencer (East) Lake.

fish were observed dying only near or within the river mouth where higher temperatures existed.

St. Lawrence River area. Commercial bait fishermen, hydro station operators, and sport fishermen reported that no mortalities of alewives occur in the hydro station and mill raceways of the Cataraqui and Gananoque Rivers which are tributaries of the St. Lawrence. But "annual mortalities" were reported to occur in the St. Lawrence River and Kingston Harbour into which the above rivers enter respectively. The annual mortality begins in late June and extends into July in this area.

A survey of the St. Lawrence River area, May 31 through June 1 of 1951, did not reveal any incidences of mortality. The raceways were

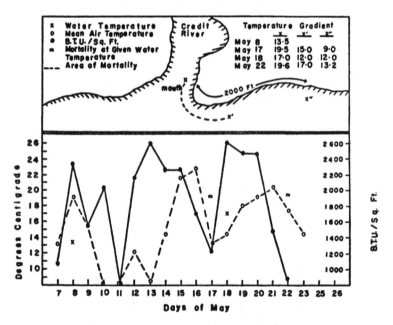

FIGURE 15.—The association of temperature gradients with occurrences of mortality of alewives observed at Port Credit in 1951.

crowded with alewives in contrast to the St. Lawrence River and Kingston Harbour where no alewives were observed. According to commercial fishermen, the latter are not occupied in numbers by the alewives until late June. Temperatures taken in the raceways and near the river mouths were as follows:

	Air	Surface	Date
Cataraqui raceway	21.0°C.	19.2°C.	May 31
Kingston Harbour	—	16.5°C.	May 31
Gananoque raceway	24.5°C.	20.5°C.	June 1
Gananoque Municipal Dock	—	15.0°C.	June 1

Temperatures recorded for the raceways were about the same magnitude as those associated with incidences of the annual mortality described above (see pages 23 and 26). That no alewives died at these high temperatures can be explained, for the alewife is able to withstand temperatures of this magnitude indefinitely under certain conditions. These conditions will be discussed below (see page 37).

Previous records. The most complete census of the annual mortality of the Lake Ontario alewife was taken by Miller (1930) in 1928. The results he obtained are plotted in figure 16. Mean air and surface water

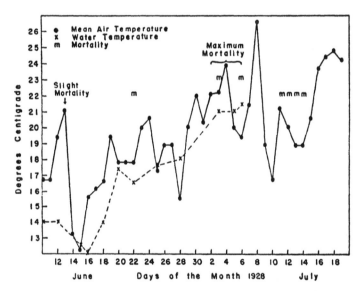

FIGURE 16.—The annual mortality of 1928 and its association with the air and surface water temperature. Data from Miller (1930).

temperatures were taken in and around Toronto Harbour during the annual mortality. The incidences of mortality noted are those in which 30 or more alewives were reported to be dying. An area extending roughly from Toronto Harbour to Port Dalhousie was investigated.

LETHAL TEMPERATURE EXPERIMENTS

Since the field observations suggested that temperature was the factor responsible for the annual mortality of alewives in Lake Ontario, tests were performed in the laboratory to find their lethal range of temperature. ("Lethal temperature" is used here to mean the temperature at which 50 per cent of the population dies on continued exposure.) The results of these experiments are presented in figures 17 and 18. Adult alewives acclimated to 10°C approach their upper lethal temperature just above 20°C.

Acclimation of adults to 15°C resulted in an expected increase to almost 23°C. The line obtained for adult alewives acclimated to 20°C shows a further increase in thermal resistance. The upper lethal temperature for 20° acclimation could not be estimated exactly but is approximately 23°C.

Young-of-the-year alewives, acclimated to 5°C, approached their upper lethal temperature at slightly less than 15°C (see figure 18). Only 60 per cent total mortality was reached at the lethal temperature of 15°C. Acclimation to 9°C resulted in a considerable increase in thermal resistance and the upper lethal temperature. Exposure to 23°C resulted in only 60 per cent total mortality, suggesting that the upper lethal temperature was only slightly less than 23°C.

DISCUSSION

Growth

The growth of both freshwater and Atlantic alewives is very rapid during the summer subsequent to hatching (see figure 4). Young of the year in Lake Ontario show a slightly faster rate of growth than those from Fletcher Creek in Nova Scotia. This is probably associated with the higher mean summer temperatures found in the Lake Ontario region.

Following their initial migration, freshwater alewives encounter an environment not greatly unlike that which they have left, while Atlantic alewives are confronted with salt water, an enviroment which is considerably different. Both the freshwater and Atlantic alewives show a decrease in growth rate following the onset of sexual maturity, but the freshwater in contrast to the saltwater environment hastens the onset of sexual maturity and the coincident decrease in growth rate is more pronounced (see figure 3).

The tendency of the onset of sexual maturity to inhibit the growth of fish is well known (see Hubbs, 1926) and has been demonstrated by Svårdson (1943), who found that a retardation of growth occurred when young guppies, *Lebistes reticulatus,* were fed oesteron and testosteron. Thus, the larger size attained by the Atlantic alewife in comparison to the freshwater alewife is chiefly related to the fact that the former experiences approximately one more year of uninhibited growth.

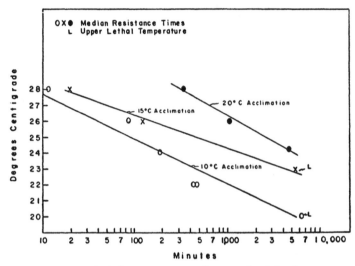

FIGURE 17.—Lethal temperature relations of adult alewives.

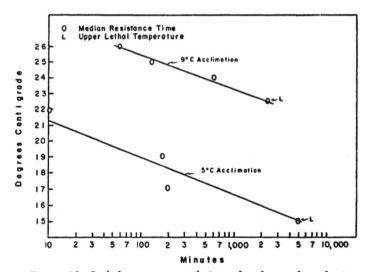

FIGURE 18.—Lethal temperature relations of under-yearling alewives.

Hoar (1952) has suggested that the slow rate of growth of the fresh-water alewife is related to the functioning of the thyroid. He examined and compared the thyroids of mature Atlantic alewives from Kenne-becasis Estuary, New Brunswick, and mature freshwater alewives from Lake Ontario. The thyroids of the Atlantic fish showed "a mild state of activity which is apparently normal for sexually maturing teleosts. . . ." Those of the freshwater alewives, in contrast, were ex-hausted. Hoar suggested that this exhaustion might be related to an insufficiency of hormone for both growth and osmotic regulation. Since the onset of maturity is not necessarily dependent on growth itself, but on those factors which also influence growth (Svårdson, 1943), excessive activity of the thyroid gland might also hasten sexual maturity. Thus, the relatively slow growth of the freshwater alewife might reflect the stress of its freshwater environment.

Form

Variation within Lake Ontario. Variation of the relative growth of the alewife within Lake Ontario is insignificant in mature fish and presumably also in immature fish. Only three characters showed differ-ences between samples taken from various parts of the lake (see figure 5). Two of these, caudal peduncle length and snout length, showed possible differences between intercepts of the relative growth lines with the ordinate.

Differences between intercepts represent the general type of racial variation within species encountered previously (Martin, 1949; Svårdson, 1950; Hart, 1952). Martin's (1949) explanation for this type of racial variation in relative growth is that the relative growth of the body parts of fish is characterized by a series of growth stanzas, each with a different growth partition constant, and that in different popu-lations the sizes attained by the individuals differed at the point of inflection where they changed from one stanza to the next. Two stages in development where such inflections commonly occur are the time of ossification and the time of sexual maturity. The relative growth pattern of the Bay of Quinte material displays inflections that coincide with the first ontogenetic event and in some cases with the second.

The inflections found in the Bay of Quinte material are summarized in Table III. The body lengths at the point of the first inflection for the characters given are grouped about a mean of 30.6 mm. and have a spread of 26 mm. to 35 mm. Body length at the second inflection varied from 60 mm. to 115 mm. The inflections at 60 mm., 68 mm., and

TABLE III

SUMMARY OF INFLECTIONS FOUND IN THE BAY OF QUINTE RELATIVE
GROWTH SERIES

Character	Slope before inflection	Length at inflection	Slope after inflection	Length at inflection	Slope after inflection
Head depth	1.58	32 mm.	1.06	—	—
Orbital length	1.56	32	0.98	115 mm.	0.82
Predorsal length	0.40	26	1.02	—	—
Body depth	1.72	30	1.17	107	0.96
C. P. length	1.58	35	1.08	—	—
Anal base	1.71	27	0.98	—	—
Pectoral length	2.05	30	1.28	81	0.99
Pelvic length	1.87	34	0.95	—	—
Caudal length	—	—	1.32	60	1.02
Dorsal length	—	—	1.22	68	0.85

81 mm. are probably not associated with the onset of sexual maturity since this event generally occurs at a standard length slightly above 100 mm. It is interesting to note, however, that body size at the point of inflection is greater for body measurements than for fin measurements.

The third character which showed a difference between samples taken in Lake Ontario was orbital length. Here, there apparently was a difference in growth partition constants rather than intercepts. This would suggest that Martin's (1949) explanation of racial variation within species might not be applicable in the case of the Lake Ontario alewife. However, the data presented here are insufficient to form a definite conclusion, and further Martin found a similar exception in the behaviour of eye diameter in his malnutrition experiments.

Comparison of Ontario and Atlantic alewives

The comparison between the Bay of Quinte and Atlantic samples given in figures 6 to 11 and summarized in Table IV shows that the relative growth of the Atlantic sample differs from that of the freshwater sample largely through differences in slope of the relative growth lines. The slopes were significantly different at the 5 per cent level or better in eight of thirteen characters compared. A significant difference in slope implies a significant difference in intercepts. Intercepts, therefore, were not tested where slopes were different. In three of the five cases, where the slope did not differ significantly, the intercept did

TABLE IV

COMPARISON OF THE GROWTH PARTITION CONSTANTS (k) DURING THE MAIN
GROWTH STANZA IN LAKE ONTARIO AND ATLANTIC ALEWIVES

(Where the slopes are significantly different the intercepts are not compared:
xx—sig. at 1 per cent level, x—sig. at 5 per cent level)

Character	Region		Stat. sig. of diff.	
	Ontario	Atlantic	Slope	Intercept
Head length	1.03	0.93	xx	—
Head depth	1.06	1.02	0	0
Orbital length	0.98	0.84	x	—
Snout length	1.02	0.96	0	xx
Predorsal length	1.02	0.99	x	—
Body depth	1.17	1.30	xx	—
C.P. length	1.08	0.98	x	—
Anal base	0.98	1.03	x	—
Dorsal base	1.03	1.01	0	xx
Pectoral length	1.28	1.12	0	xx
Pelvic length	0.95	1.00	0	0
Caudal length	1.32	1.17	x	—
Dorsal length	1.22	1.11	x	—

differ, showing the customary type of racial variation referred to on page 30.

Perhaps the preponderance of differences in the values of k between salt- and freshwater samples indicates that as the alewife grows older the freshwater environment may become more and more unsuitable for it, progressively eliminating more and more individuals which represent a certain part of the range of morphometric variation. If such selection were occurring, it would be expected that the variability within the population with respect to any character influenced by such selection would become less with increasing size. Although there are apparently few series of data extensive enough to afford a reliable statistical test, the variance about a relative growth plot is considered to remain constant. Martin's (1949) best documented example is from his investigation of a hatchery stock of brook trout, *Salvelinus fontinalis*. In these he observed that differences in the relative size of the body parts found in the original sorting of the lots were maintained to the same relative degree throughout subsequent growth.

To determine the degree of variability in body parts in the Lake Ontario alewife throughout life, the dispersion of the head length was determined at three size ranges in the Lake Ontario sample.[1] One hundred specimens were measured over each of the size ranges 37.5–43.9 mm., 59.3–71.5 mm., and 120.0–130.0 mm. The ranges (two standard deviations) for the head lengths of each size group were calculated and then plotted in figure 19 at the mean standard length of each respective size group.

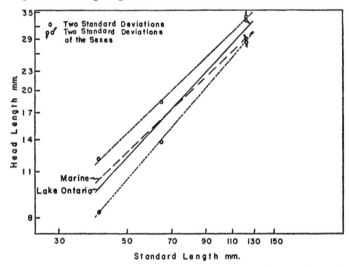

FIGURE 19.—Relative variability of head length throughout life in the Lake Ontario alewife and its relation to the difference in slope of mean head length in relation to body length between the Lake Ontario and the Atlantic alewife.

The data on the variability of the head length as displayed in figure 19 together with the position of the relative growth line for the Atlantic alewife strongly suggest that the fish with shorter heads are removed from the population at a rate faster than the general mortality rate. Selection is suggested by the reduction in variability in the upper size range while the slope of the mean head length for the Atlantic alewives essentially parallels the upper limit of variation but converges with the lower limit.

It seems probable that the other differences in slope found between the freshwater and Atlantic samples can be explained in a similar

[1]Most of the specimens were obtained from Bay of Quinte samples. However, some from the Port Credit and Frenchman Bay areas were also used in the smallest and medium size ranges. This is not thought to introduce any error into the degree of variance since it has been shown that specimens from these areas and the Bay of Quinte are homogeneous for this character.

fashion since there are correlations between head length and other body part measurements. These correlations are presented in Table V, which gives the calculated values for their coefficients of regression. The following method was used to obtain these figures and values. The head lengths of the ten specimens in each logarithmic group of the Bay of Quinte sample (see page 3), were rated from small to large. Then the smallest head lengths in each group were summed and the next smallest and so on. Subsequently, the other measurements to be correlated were arranged in the same order as their respective head lengths and summed. The figures and values obtained, and presented in Table V, represent only those segments of the relative growth lines

TABLE V

SLOPE OF REGRESSION LINE OF BODY MEASUREMENTS ON HEAD LENGTH

(b: values of the slope of the regression line of various body measurements on head length. These characters are those which showed significant differences in slope between the marine and fresh water samples. A t value of 2.31 or greater is significant at the 5 per cent level)

Character	b	t
Orbital length	0.291	1.31
Predorsal length	0.293	3.46
Body depth	0.399	2.75
C.P. length	0.395	2.03
Anal base	0.241	1.59
Caudal length	0.358	2.58
Dorsal length	0.554	3.81

of the characters which were compared statistically. Four of the seven characters show significant regressions by the method used.

Differences in proportional parts. Pritchard (1929) described differences in the morphometry of the freshwater and Atlantic adult alewives expressed as differences in body proportions. He measured ten specimens of 130 mm. to 176 mm. standard length from the Port Credit area and ten Atlantic specimens from Halifax Harbour, 239 mm. to 279 mm. in length. He characterized the Atlantic specimens as having longer fins, much shorter heads, and smaller eyes. He also presented his original data, together with the means of measurements taken. These means are plotted in figure 6 and figures 8 to 11. The data of Pritchard and that of the author show a close agreement in the case of head length, body length, pectoral length, and pelvic length. Differences in regard to other characters appear to be related to differences of technique and to the small number of fish measured by Pritchard. The statement by Pritchard that the Atlantic alewife has longer fins does not agree with the findings of the author (see figures 10 and 11), but the data are insufficient for accurate comparison.

Pritchard qualified his statement that the Atlantic alewives had much shorter heads by suggesting that this difference, ". . . might be due only to their larger size." Figure 6 shows that this qualification is correct. If Pritchard's Atlantic specimens had been of the same body size as his freshwater specimens, the difference between their head length would have been much less. If he had measured fish at about 75 mm. standard length, the differences would not exist. Below this length, the situation would be reversed and the freshwater alewives would have the shorter heads.

Differences in vertebral counts. The size effect shown above in the case of head length is also evident when correlations of vertebral counts are made with other characters. It is a general rule that a population of fish which have fewer meristic parts than another population of the same species, also have longer body parts, such as heads, eyes, and fins. However, this correlation does not necessarily hold throughout the entire life span of the fish. When a comparison is made with adult freshwater and Atlantic alewives such a correlation is evident in the case of head length (figure 6), orbital length (figure 7), paired fins (figure 10), and caudal length (figure 11). However, the correlation is clearly reversed for head length and orbital length when standard lengths somewhat smaller than those of adult fish are considered.

Since the establishment of meristic parts occurs early in life, the saltwater environment of the anadromous Atlantic alewife cannot be regarded as responsible for the statistically significant differences noted between the vertebral counts of the freshwater and Atlantic samples. Such differences have long been known to be associated with temperature gradients. Also, alteration of the developmental rate of fishes has been shown to result in differences in the number of meristic parts (see Martin, 1949, p. 5). Therefore, it is possible that the differences in temperature between the two regions where development of the samples took place (Lake Ontario and Nova Scotia) was responsible for the difference in vertebral counts.

Mortalities

The alewife, during its winter occupation of deep water, is exposed to a constant low level of temperature (about 3° to 5°C). Thus it is acclimated to a low temperature when it reaches inshore waters during the spring shoreward migration. This was corroborated by the low carrying temperature (8.5°C) necessary to transport alewives successfully to the laboratory in the early summer. The spring invasion of inshore shoal waters brings the alewife into a region where temperatures approach 20°C for various periods of time. During this spring invasion the alewife sometimes enters temperature gradients such as those shown in figures 12 and 15. Entrance of the fish into the gradients is probably related to the directive action of temperature. Collins (1952) has shown that when the Atlantic alewife and the related Atlantic species *Pomolobus aestivalis* during their spring migration upstream, were presented with a choice of waters having different temperatures, they showed a preference for the warmer water. This occurred when the temperature difference between the waters continuously exceeded 0.5°C and water temperatures varied from 11.1°C to 22.3°C when the experiments were performed.

The Lake Ontario alewives die over a certain range of temperature within the above gradients. In the Bay of Quinte region this range extended from a little less than 18.5°C through 19°C (see figure 12), and in the Port Credit area from 17°C through 19.6°C (see figure 15). A comparison of these ranges with the lethal temperature relationships obtained for the alewife in the laboratory is shown in figure 20. This comparison indicates an accord between data obtained in the field

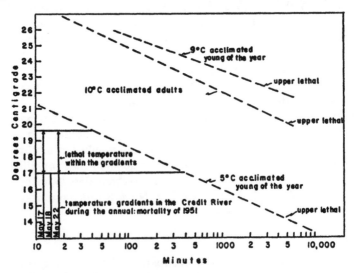

FIGURE 20.—The relation of lethal temperatures found in the laboratory to water temperatures observed at the time of mortalities.

and in the laboratory. It shows that the range of temperature in the temperature gradients within which the alewives died corresponds to a respective range of thermal resistance extending from 45 minutes at the highest point (19.6°C) to 430 minutes at the lowest (17°C). That is, 50 per cent of the young of the year (and presumably juveniles and adult alewives also), acclimated to 5°C, upon entering this range of the temperature gradients (19.6°C through 17.0°C), would die sometime within 45 to 430 minutes. Figure 20 shows that acclimation to higher temperatures once inshore would preclude further death of alewives in the temperature gradients since the upper lethal temperatures of fish acclimated to 9°C and 10°C are above 19.6°C, the highest temperature of the gradients.

The above explanation of the cause of the annual mortality is not consistent with the later and more severe incidents which occur in late June through mid-July. However, two things suggest that temperature is the lethal factor in the later and more severe mortalities as in the early incidences. There is a steady increase in the extent of mortality which is coincident with an increase in temperature (see figure 16). The mortality ceases shortly after the maximum surface temperature is reached around mid-July. The mortalities occur during periods of calm water and of bright sunshine (high temperature). Miller (1930) states concerning the annual mortality of 1928, ". . . we have noticed in many instances that a large mortality has followed upon a day of complete calm and high temperature."

There are two apparent exceptions to the action of lethal temperature as a causative agent in the annual mortality. Alewives died in the offshore surface water of the lake as well as in the inshore shoal water and no mortalities have been reported in the raceways of the Cataraqui and Gananoque Rivers (see p. 26). In the offshore surface water, death from lethal temperature probably results when the alewife schools invade the uppermost level of the surface water during periods of calm water and bright sunshine. It is suggested here that the lack of mortalities in the raceways under temperature conditions which produced death elsewhere (see pp. 23 and 27), is possibly related to the shallowness of the water through which these fish migrate and that they are early arrivals in the area. Their migration route begins in the lake proper and lies more along shoal water (Bay of Quinte and St. Lawrence River) than that of the fish which migrate directly from the depths of the lake proper towards its shore. The temperatures of Kingston Harbour (16.5°C) and the St. Lawrence River (15°C) are still relatively low when the alewives migrate to the raceways. Presumably, there is sufficient time for them to gain additional heat tolerance so that the high temperatures (19.2°C and 20.5°C) of the raceways are no longer lethal. In contrast, the annual mortality begins about late June or early July in Kingston Harbour and the St. Lawrence River, presumably after their temperatures have increased. The major portion of the alewives migrating shoreward enter the harbour and river at this time; simultaneously the raceways are deserted.

Other mortalities of alewives have been reported throughout the year which are not associated with the annual mortality. Those occurring during the early spring are probably related to the fragility of the alewife when acclimated to low temperature. This fragility is demonstrated by a susceptibility to loss of equilibrium and to attack by *Saprolegnia*. Alewives were observed during May 1951 to surface and thresh about in the wake of power boats. According to commercial fishermen this action ceases during the summer, and dynamite caps exploded in the midst of alewife schools had little effect during late July 1950. However, Miller (1930) reported that commercial fishermen caused alewives to perform circus movements and to die in early July by pounding two rocks together under water in the vicinity of the fish. *Saprolegnia* had often been reported on dead and dying alewives (Miller, 1930; Rathbun, 1895). It was prevalent during the spring of 1950 but was not noticed during the subsequent summer. Its presence in the laboratory coincided with low acclimation temperatures (5°C and 10°C).

No explanation can be suggested for the mortalities of young-of-the-year alewives during the fall and winter months in Lake Ontario. Breder and Negrelli (1936) reported a mortality of alewives in Kensico Reservoir during the late fall and winter of 1935. The authors suggested that the mortality was associated ". . . with that great group of periodical fluctuations of animal populations. . . ."

Briefly, the annual mortality of the alewife appears due to the alewife's inability to acclimate rapidly to rising or fluctuating temperatures. Early incidences of the mortality follow when the alewife invades the warm spring shoal waters while acclimated to the cold temperatures of the lake bottom. Later and more extensive incidences, although not fully explained here, are also probably associated with lethal temperatures.

Mortalities other than the annual mortality have not been explained. Those occurring during the winter and early spring presumably are associated with the fragility of the alewife when acclimated to low temperatures.

Differential mortality

Since the annual mortality seems to be brought about by physical events there is a possibility that fish of certain bodily form may be preferentially killed by the operation of the lethal factor. It appears likely, since this mortality is a feature of the landlocked group, that a loss of osmotic control acting as an accessory factor to lethal temperature brings about the death of the freshwater alewife.

The sites of osmotic exchange of the alewife with the external environment are chiefly its gills and oral membranes and these sites are located in the head. Keys (1931) found that larger-headed killifish, *Fundulus parvipinnis*, had a greater ability to withstand adverse conditions, that is, decrease of salinity and asphyxiation. He suggested this ability was related to the relatively greater area of gill surface of the larger-headed fish. If large-headed fish have an advantage in osmotic regulation, then the progressive loss of the smaller-headed representatives is easily explained. The greater ability of larger-headed fish to survive the mortality is possibly related to "chloride secreting" cells found in the gills of fish. Keys and Willmer (1932) found large ovoid cells at the base of the gill leaflets of several fishes and suggested that they performed the special function of chloride secretion (inward or outward) to maintain osmotic balance. Krogh (1939) has also suggested that the entire respiratory epithilium might be secretory. Thus, fish with larger heads probably have a greater number of chloride secreting cells and perhaps a larger respiratory secretory surface with which to control the osmotic balance of their bodies. These larger-headed fish are not selected out of the Lake Ontario alewife population as is shown in figure 18.

SUMMARY

1. The ages of 168 freshwater alewives from the Bay of Quinte region of Lake Ontario and 141 Atlantic alewives from Nova Scotia and New Brunswick were determined by the scale method. A comparison of the author's data for the freshwater alewife and for the Atlantic alewife showed that the latter had a more rapid rate of growth, as has also been suggested by other authors. Both the freshwater and Atlantic alewives displayed an early rapid rate of growth followed by a decrease in growth rate coincident with the onset of sexual maturity. It is suggested that, in contrast to the saltwater environment, the freshwater environment hastens the onset of sexual maturity with its growth inhibiting tendencies. The Lake Ontario alewife matures about one year earlier (age groups II ♂ and III ♀) than the Atlantic alewife (age

groups III ♂ and IV ♀). Thus, it is concluded that the larger size attained by the Atlantic alewife is related to the fact that it experiences approximately one more year of uninhibited growth than the freshwater alewife.

Hoar (1952) has proposed that the slow growth of the Lake Ontario alewife is related to excessive stimulation of the thyroid mechanism which causes an insufficiency of hormone for both growth and osmotic regulation. This is thought to be in agrement with the above findings since excessive stimulation of the thyroid gland might also hasten sexual maturity and thus retard growth.

2. Comparison of samples of adult alewives taken from various areas of Lake Ontario showed that variation in the relative growth of the alewife within Lake Ontario was essentially small. Of the thirteen measurements made, only three showed slight differences between samples. In two, the slopes of the relative growth lines of the samples were parallel, but possibly with different intercepts which is the general type of racial variation encountered previously (Martin, 1949; Svårdson, 1950; and Hart, 1952). This type of variation was explained by Martin, who showed that in different populations the sizes attained by individuals differed at the point of inflection where they changed from one relative growth stanza to the next.

In the case of the Bay of Quinte sample inflections were found to occur at a mean standard length of 30.6 mm. and over a range of 60 mm. to 115 mm. The early inflection may be coincident with ossification while two later inflections at 107 mm. and 115 mm. probably are related to the onset of sexual maturity. No explanation is offered for those later inflections occurring at 60 mm., 68 mm., and 81 mm., which are all below the standard length (about 100 mm.) at which sexual maturity occurs. Only the later inflections were found in the Atlantic alewife sample. These paralleled those of the Bay of Quinte sample, but occurred at a larger body size.

The other measurement showing a difference between samples taken from Lake Ontario suggested a possible difference in slope. This followed the pattern of variation shown between the Bay of Quinte and Atlantic samples, where there were also differences between slopes. In eight of the thirteen characters compared, differences of slopes were statistically significant between samples. Three of the remaining five characters showed differences between intercepts as described above.

The variability of head length of the Lake Ontario alewife was measured at three positions along the relative growth line for head length. The standard deviation of head length decreased progressively from small to large standard lengths while the slope of head length for the Atlantic alewife roughly paralleled a line two standard deviations above the mean but converged with the corresponding limit drawn below the mean. Regression values between head length and the other seven characters showing differences between slope were calculated. Values were obtained ranging from 0.29 through 0.55 for coefficients of regression.

It is concluded that the differences in body form between the freshwater alewife and the Atlantic alewife are chiefly the result of differential selection of the freshwater alewife by its environment.

The mean difference betwen the number of vertebrae of the Lake Ontario and Atlantic sample from Nova Scotia was found to be statistically significant. The lesser number of vertebrae of the Lake Ontario alewife is attributed by the author to the higher developmental temperatures of the Lake Ontario region.

The general rule, that a population of fish which have fewer meristic parts than another population of the same species also have longer body parts, does not necessarily hold throughout the entire life span of the alewife.

3. During the alewife's spring invasion of shoal water in Lake Ontario it sometimes enters temperature gradients that have maximum surface temperatures approaching 20°C. Alewives were observed to die principally within the upper ranges of temperature of these gradients. A correlation between the gradual increase in surface temperature from spring to summer with a parallel increase in the incidences of the mortality was also shown.

Adult alewives acclimated to 10°C, 15°C, and 20°C approached their upper lethal temperatures (that temperature at which only 50 per cent of the sample dies on continued exposure) at just above 20°C, just below 23°C, and about 23°C respectively. Young-of-the-year alewives acclimated to 5°C and 9°C had upper lethal temperatures slightly below 15°C and 23°C.

Comparison of the range of lethal temperature obtained in the laboratory with the lethal ranges of the temperature gradients observed in the field showed that the early incidences of the mortality are related to the entrance of alewives into warm shoal waters during the spring while acclimated to the low temperatures of the lake's depths. Later and more severe incidences are believed also to be related to lethal temperature although no support for observations made in the field was found in the laboratory experiments.

Inadequacy of osmotic regulation by the Lake Ontario alewife is thought to be acting as an accessory factor to a lethal factor, probably temperature, in the selection of small-headed fish out of the Lake Ontario alewife population. It is suggested that fish with relatively large heads have a greater number of chloride secreting cells within their gills with which to maintain the osmotic balance of their bodies.

REFERENCES

BREDER, C. M., Jr., and R. F. NEGRELLI. 1936. The winter movements of the landlocked alewife, *Pomolobus pseudoharengus* (Wilson). Zoologica, 21(13): 165–75.

BRETT, J. R. 1952. Temperature tolerance in young Pacific salmon, genus *Oncorhynchus*. Jour. Fish. Res. Bd. Canada, 9(6): 265–323.

COLLINS, GERALD B. 1952. Factors influencing the orientation of migrating anadromous fishes. U.S. Dept. Int., F.W.S., Fish. Bull. 73: 373–96.

DELURY, D. G. 1951. On the planning of experiments for the estimation of fish populations. Jour. Fish. Res. Bd. Canada, 8(4): 281–307.

FRY, F. E. J. 1951. A splashless tank for transporting fish. Can. Fish. Cult., no. 10: 15–16.

FRY, F. E. J., J. S. HART and K. F. WALKER. 1946. Lethal temperature relations for a sample of young speckled trout, *Salvelinus fontinalis*. Univ. Toronto Studies, Biol. Ser. no. 54: 9–35; Pub. Ont. Fish. Res. Lab. no. 66.

HART, J. S. 1952. Geographical variations of some physiological and morphological characters in certain freshwater fish. Univ. Toronto Studies, Biol. Ser. no. 60: 1–63; Pub. Ont. Fish. Res. Lab. no. 72.

HOAR, WILLIAM S. 1952. Thyroid function in some anadromous and landlocked teleosts. Trans. Roy. Soc. Canada, Ser. 3, vol. 46, Sect. 5: 39–53.

HUBBS, CARL L. 1926. The structural consequences of modifications of the developmental rate in fishes considered in reference to certain problems of evolution. Am. Nat., 60: 57–81.

HUNTSMAN, A. G. 1918. The growth of scales in fishes. Trans. Roy. Can. Inst., Part 1(27): 61–101.

—— 1938. Overexertion as cause of death of captured fish. Science, 87(2269): 577–8.

HUXLEY, JULIAN S. 1932. Problems of relative growth. London: Methuen. Pp. i–xix; 1–276.

HUXLEY, JULIAN S., J. NEEDHAM and I. M. LERNER. 1941. Terminology of relative growth rates. Nature, 148: 225.

KEYS, ANCEL B. 1931. A study of the selective action of decreased salinity and of asphyxiation on the Pacific killifish, *Fundulus parvipinnis*. Bull. Scripps Inst. Oceanog., Tech. Ser., 2(12): 417–90.

KEYS, ANCEL B., and E. N. WILLMER. 1932. "Chloride secreting cells" in the gills of fishes, with special reference to the common eel. Jour. Physiology, 76(3): 368–78.

KROGH, AUGUST. 1939. Osmotic regulation of aquatic animals. Cambridge Univ. Press. Pp. 1–242.

LEIM, A. H. 1924. Life history of the shad. Contrs. Can. Biol., New Ser., 2(11): 161–284.

MARTIN, W. R. 1949. The mechanics of environmental control of body form in fishes. Univ. Toronto Studies, Biol. Ser. no. 58: 1–91; Pub. Ont. Fish. Res. Lab. no. 70.

MILLER, G. W. 1930. The mortality of alewives (*Pomolobus pseudoharengus* (Wilson)) in Lake Ontario. Ont. Fish. Res. Lab., Manuscripts 1–47.

ODELL, T. T. 1934. The life history and ecological relationships of the alewife (*Pomolobus pseudoharengus* (Wilson)) in Seneca Lake, New York. Trans. Amer. Fish. Soc., 64: 118–26.

ODELL, T. T., and STEPHEN W. EATON. 1940. The rate of growth of the alewife (*Pomolobus pseudoharengus* (Wilson)) in several lakes of New York. Proc. N. E. Fish. Cult. Meeting at Trinity College, Hartford, Conn. on April 5 and 6, 1940: 3–4 (multigraphed).

PRITCHARD, ANDREW L. 1929. The alewife (*Pomolobus pseudoharengus*) in Lake Ontario. Univ. Toronto Studies, Biol. Ser. no. 33: 37–54; Pub. Ont. Fish. Res. Lab. no. 38.

RATHBUN, RICHARD. 1895. Report on the inquiry respecting food fishes and the fishing-grounds. U.S. Comm. Fish & Fisheries, Rept. of the Commissioner for the year ending June 30, 1893, Part 19: 17–51.

ROUNSEFELL, GEORGE A. and LOUIS STRINGER. 1945. Restoration and management of the New England alewife fisheries with special reference to Maine. Trans. Amer. Fish. Soc., 73: 394–424.

SMITH, HUGH M. 1892. Report on an investigation of the fisheries of Lake Ontario. Bull. U.S. Fish. Comm., 1890, 10: 177–215.

SVÄRDSON, GUNNAR. 1943. Studien über den Zusammenhang zwischen Geschlechtsreife und Wachstum bei *Lebistes*. Kungl. Lantbruksstyrelsen Meddelanden från Statens undersöknings–och försöksanstalt för sötvattenfisket., Nr. 21: 1–48.

——— 1950. The coregonid problem II. Morphology of two coregonid species in different environments. Inst. Fresh-water Research Drottningholm, Rept. no. 31: 151–62.

THOMPSON, D. H. 1934. Relative growth in *Polyodon*. Ill. Nat. Hist. Survey, Biol. Notes no. 2: 1–8.

Lightning Source UK Ltd.
Milton Keynes UK
UKHW012358200722
406167UK00001B/308